Thomas Tugby

Tugby's Guide to Niagara Falls

A Complete Guide to all the Points of Interest Around and in the Immediate

Neighbourhood

Thomas Tugby

Tugby's Guide to Niagara Falls
*A Complete Guide to all the Points of Interest Around and in the Immediate
Neighbourhood*

ISBN/EAN: 9783743442696

Manufactured in Europe, USA, Canada, Australia, Japa

Cover: Foto ©berggeist007 / pixelio.de

Manufactured and distributed by brebook publishing software
(www.brebook.com)

Thomas Tugby

Tugby's Guide to Niagara Falls

WINTER WONDERS IN THE PARK.

WHIRLPOOL RAPIDS RAILWAY SUSPENSION BRIDGE AND CANTILEVER BRIDGE

SWIMMING THE WHIRLPOOL RAPIDS

INDIAN FAMILY

NEW SUSPENSION BRIDGE AND FALLS

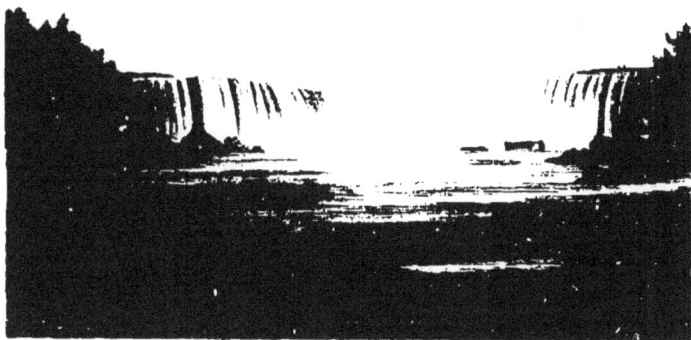

HORSE SHOE FALL FROM CANADA

AMERICAN FALL AND VILLAGE OF NIAGARA FALLS

HORSESHOE FALL AND THE MAN   TABLE ROCK

BELOW PROSPECT PARK

INTERIOR RAILWAY SUSPENSION BRIDGE, ROEBLING [...]

AMERICAN FALL FROM BELOW

BRIDGE TO FIRST SISTER ISLAND

BRIDGE TO SECOND SISTER ISLAND

SPARE THE TREES AND SHRUBS

BRIDGE TO THIRD SISTER ISLAND

SISTER ISLAND

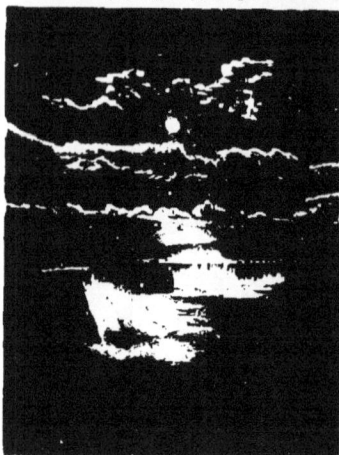

GENERAL VIEW FROM NEW SUSPENSION BRIDGE   RAPIDS FROM THIRD SISTER ISLAND

TUGBY'S
ILLUSTRATED GUIDE.

TO
NIAGARA FALLS.

TUGBY'S GUIDE TO NIAGARA FALLS

being

a complete guide to all the points of interest around

and in the immediate neighbourhood.

EMBELLISHED WITH VIEWS;

copied from photographs

made especially for this work.

PUBLISHED

by

THOMAS TUGBY, BRIDGE ST., NIAGARA FALLS, N. Y

A resident of Niagara Falls for thirty years, and proprietor of
Tugby's Mammoth Bazaar.

To the visitor!

~~~~~

In arranging this work, great care has been taken to make it in every respect, correct, and worthy of the attention of the visitor, and to render it both concise and comprehensive, combining all the useful features of my other publication on Niagara, with numerous improvements, calculated to adapt it to the present demands. With these many features of excellence, it is confidently hoped that this work will be appreciated, as a valuable assistant to the visitor, and a ready guide and advisor, under every difficulty which may beset the tourist at Niagara Falls.

The author.

INDEX.

by

D. C.

Tugby's guide to Niagara Falls.

Introduction.

Nature has many water-falls and cataracts, but only one Niagara. Niagara Falls may be justly classed among the wonders of the world. They are the pride of America. Their grandeur magnitude, and magnificence, are well known to the civilized world. Ever since the discovery of this wonderful cataract, thousands have flocked hither from all countries, to gaze with feelings of the deepest solemnity on the tumultuous fall of waters, and to adore the power and majesty of the Almighty as these are exhibited and realized amid the sublime scenery of this stupendous water-fall. The power of the Almighty is here more grandly exhibited than in any other scene on earth. The Falls cannot be described, there is too much sublimity, majesty and over whelming grandeur for mortal to comprehend or explain. The great features of Niagara are ever the same, but their individual expression is continually changing. With every season, with every sunbeam, with every shade, they assume a different appearance, inspire fresh interest, and compel new admiration. No place on the civilized earth offers such attractions and inducements to visitors. They can be never fully known, except to those who see them, from the utter impossibility of describing such a scene. When nature can be expressed by color, then there will be some hope of imparting a faint idea of it — but until that can be done, Niagara must remain undescribed. At first sight, visitors are sometimes disappointed. Either their expectations have been raised too high, or the scene surpasses what they had anticipated. The second view is frequently more impressive than the first.

The longer the visitor tarries the more he enjoys a appreciates; we think the instance is not on record where t visitor, having any mind to appreciate overwhelming grandeur, h been disappointed after a few days at Niagara. The cataract formed by the precipitous descent of the Niagara-River, down a led of rocks, more than one hundred and sixty feet of perpendicu height, into an abyss, or basin below of unknown depth.

The Niagara-River is the outlet by which the vast surplus the waters of Lake-Superior, Michigan, Huron, and Erie is pass into Lake-Ontario, thence by the river and Gulf St. Lawrence ir the ocean.

This river forms the boundary-line between the American a Canadian domains, and divides the Horse-shoe Fall between t two countries.

The Cataract is situated in latitude 43° 6" north, and longitu 2° 6" west, from Washington, twenty-two miles from Lake Erie a fourteen from Lake Ontario. In the following pages we sh attempt to guide the traveller to all the points of interest whe the best views may be obtained, and thereafter point out to h the spots of peculiar interest in their neighbourhood. Let suppose, now, reader, you have just reached the village of Niaga: Falls on the American-side; that you have just alighted from t train, and that you do not want to ride or go to a hotel. is superfluous to give you minute directions how to procee follow the carriages and stages and you will be right. Visite generally take an omnibus from the depot, to a hotel previou selected. All the principal hotels send an omnibus to every tra The distances to the different points are given, so that the tout may either walk or ride as he may elect. It would, perhaps be as w to state that the distances are too great to walk to all points of inter€

The street between the Cataract and International Hotels le₂ to the river where it is spanned by

Goat Island Bridge.

The bridge is about 40 rods above the Falls. This eleg: and substantial structure, was erected by Mess™ Porter, ₁ proprietors of Goat-Island. It is on the plan of the Whipple iron arched bridge, is 360 feet long, having four arches of ninety feet span each. Their width is twenty seven feet. The first bridge that was thrown across these turbulent waters was constructed at the head of Goat-Island in 1817, it was carried away by the ice in the following spring, and was succeeded by another built in 1818 on the site of the present bridge. This was repaired in 1839, again, in 1849. The present bridge was erected in 1856. The inquiry is often made, how was it ever constructed over such a tremendous rapid? The difficulties attending its construction were overcome in the following manner: A massive abutment of timber was built at the water's edge, from which were projected enormously long and heavy beams of timber. These beams were secured on the land side by heavy loads of stone; their outer ends were rendered steady by means of stilts or legs let down from them and thrust into the bottom of the river. A platform was thrown over the projection, along which a strong framework of timber, filled solidly with store, was carried and sunk into the river. To this pier the first permanent portion of the bridge was fixed; then commencing from the extremity, beams were run out and a second pier similarly formed; and so on till the bridge was completed.

This is one of the finest points of view from which to observe the

Rapids above the Falls.

Here the first perceptions of power and grandeur begin to awaken in our minds. The noble river is seen hurrying on towards its final leap; as we stand upon the bridge, looking down upon the rushing flood of water, that seems as if it would sweep away our frail standing-ground and hurl us over the dread precipice, whose rounded edge is but a few yards further down, we begin, though feebly as yet, to realize the immensity of this far-famed cataract. The fall of the river from the head of the rapids (¾ of a mile above) to the edge of the precipice is nearly 60 feet, and increases in velocity, from seven to thirty miles an hour. The tumultuous madness of the waters, hurling and foaming in wayward billows and breakers down the descent, as if fretting with impatience, is a fine contrast to the uniform magnificent sweep with which at length they rush into the thundering flood below. Midway between the bridge and precipice will be seen Avery's Rock (see description of Avery on log).

We will now pass over the bridge to

Bath-Island.

A small Island of about one acre in extent, upon which is the Niagara Falls paper mills, said to be one of the largest paper manufacturing establishments in the United-States. A little higher up are two small Islands, called "Ship" and "Brig" Island, from their supposed resemblance to that particular kind of craft. The former is frequently called Lover's-Retreat. Looking down the river, we see several small Islets, most of which are more or less connected with thrilling incidents; for graphic details of which we refer the traveller to the carriage-drivers and guides, who are learned in local tradition.

We will now pass over another bridge similar to the one just passed (but smaller), to

Goat-Island.

Goat-Island contains 62 acres; is a little over a mile in circumference, and heavily timbered. In 1770, a man by the name of Stedman placed some goats here to pasture, hence the name. It was originally called Iris Island, from the number of beautiful rainbows that are so frequently seen near it. Goat Island was visited long before the Bridges were constructed, but the visitors were not numerous, the risk being very great. The dates 1771, 1772, 1779, under the names of several strangers, were found cut in beechtrees near the Horse-shoe Fall.

In 1814, General and Judge Porter bought of Samuel Sherwood a paper called a *Float,* given by the state as pay for military services rendered, authorizing the bearer to locate 200 acres of land on any of the unsold or unappropriated land belonging to the State. Part of this they located on Goat and other adjacent islands, immediately above and adjoining the Great Falls, their patent bearing date 1816 and signed by Daniel D. Tompkins as Governor, and Martin Van Buren as Attorney-General of New York. An early record says the Island once contained 250 acres of land.

In approaching the Island we ascend the hill and take the road to the right (five minutes walk)which leads to

Luna-Island.

This little Island, adjacent to Goat Island is connected with it by a foot bridge over the stream that forms the Centre Fall; this stream, though a mere ribbon of white foam when seen from a short distance, in contrast with the other F , is by no means unworthy of notice. It is one hundred , and is a very graceful sheet of water. From Luna Island (which is so called because it is the best point from which to view the beautiful lunar-bow) a view of the river below the falls, the Inclined Railway, the Cave of the Winds, the two suspension Bridges in the distance, the American and Centre Falls, may be seen to advantage, the visitor being located at the edge of the precipice of the American Fall. This view is thought by many to be unsurpassed.

It has often been remarked by strangers that this Island trembles, which is undoubtedly true; but the impression is heightened by imagination.

It was while climbing over the rocks directly under this Island, that Dr. Hungerford, of Troy, N. Y., was killed in the Spring of 1839, by the crumbling of a portion of the rock from above. This is the only accident that ever occurred at Niagara by the falling of rock.

Returning to Goat Island, we proceed a few yards to

Biddle Stairs.

They were erected in 1829, for the purpose of enabling visitors to descend to the Cave of the Winds, and were named after Mr. Biddle of Philadelphia, who contributed towards their erection. The stairs are 80 feet high. The total distance from top of bank to bottom is 185 feet. Number of steps 132. Here are dressing rooms for those who wish to go into the far-famed

Cave of the Winds.

The appellation by which it is known is entirely appropriate. Width of cave 100 feet, diameter 60 feet, height 100 feet. It is necessary to put on water-proof dresses and obtain a guide. (See admission fees). The cave is much visited by ladies as well as gentlemen. It was formed by the action of the water on the soft substratum of the precipice, which has been washed away and the lime-stone rock left arching overhead 30 feet beyond the base. In front the transparent Fall forms a beautiful curtain. In consequence of the tremendous pressure on the atmosphere, the cave is filled with perpetual storms, and the war of conflicting elements is quite chaotic. A beautiful rainbow, quite circular in form quivers amid the driving spray when the sun shines. Along the floor of this remarkable cavern the spray is hurled with great violence, so that is strikes the walls and curls upwards along the roof thus causing the turmoil which has procured for this place the title which it bears.

Here you may walk out on bridges and platforms, directly in front of the Falls about 40 feet distant. With the Falls pouring down at your feet, you are in the midst of heavy spray, and are almost deafened with the roar and general tumult around you: Truly this is a scene never to be forgotten. No tourist should miss seeing this remarkable phenomenon.

Near the Biddle stairs the celebrated

Sam Patch

made two successful leaps, in 1829. A ladder was placed at the foot of the rock, and fastened with ropes in such a manner that the top projected over the water. A platform was then laid on the top of the ladder from which he jumped into the river, a distance of 97 feet. Not content with the achievement, he afterwards made a higher leap of Genesee Falls, Rochester, where he was killed. After ascending the stairs we proceed a little further, walk down a few steps, and cross a little bridge to what was once

Terrapin Tower.

This tower occupied a singular and awful position. A few scattered rocks lie on the very Brink of the Fall, seeming as if unable to maintain their position against the tremendous rush of water; upon these rocks the tower was built; it was erected in 1833 by Judge Porter. A few years ago it was removed being considered unsafe.

Here we obtain the most magnificent view that can be conceived: the rapids above rolling tumultuously towards you — the green water of the mighty Fall at your feet — below you the hissing caldron of spray, and the river with its steep bank beyond; in fact, the whole range of the Falls themselves, and the world of raging waters around them, are seen from this commanding point of view.

Passing on along the edge of the Rapids (5 minutes walk) we come to the

Three Sister Islands.

These Islands are now connected to Goat Island by three beautiful foot bridges, from which the best view of the rapids is to be obtained, (see narrow escapes).

Continuing our way a short distance along the rapids we come to the

Head of Goat Island.

Here we view the broad and placid river above, spread out in a beautiful wide sheet of water. In the distance we see Navy Island, celebrated in the history of border warfare, the site of old Fort Schlosser, on the American side, and the town of Chippewa on the Canada shore.

You notice the current is rapid but not broken here, and sets directly from the shore to this spot. Here is where it is supposed the first white man ever stood upon Goat Island. Israel Putnam, in 1755, while on a campaign against Fort Niagara, at the mouth of the river, visited this place and made a trip to the head of this island and returned. The Indians seem to have crossed occasionally, as traces of their graves have been found here.

Leaving here we return to the bridge, having made the entire circuit of the Island. We have now made a journey of a little over 2 miles.

We will now recross the bridges and hasten down the banks of the river (about 150 yards to

Prospect Park.

The Park contains about 10 acres, and it embraces fine views of the American, Centre and Horse shoe Fall. This is indeed a sight worth coming many miles to see. Here, at one wide sweep, we behold Niagara stretching from the American to the Canada side in magnificent prospective. Between, as if in the grasp of the cataracts, Goat Island seems to hang precariously above the abyss. The scene is certainly one of the grandest of terrestrial nature.

Turning from the Falls, we come to the

Inclined Rail-way

down which visitors are carried in cars, worked by water power, to the edge of the river below the Falls, where a sublime view of the American Fall can be had from its base. By the side of this inclined railway is a stairway, by which those who prefer it, can walk down instead of riding in the cars (Number of steps 290.)

Those who prefer it, can now take the ferry boat to the Canada shore, a voyage of some ten minutes duration which is perfectly safe, not one accident having happened in fifty years;

some consider the view from the ferry boat the grandest of all. The depth of the river is about 200 feet.

Now reader, we have viewed the Falls from all the principal points on the American side, and those who do not take the boat to Canada will return with us to the park above.

If the preceding tour has been done on foot, those who have followed us thus far will need a little rest, after which, we suggest a carriage be now engaged, then we will proceed to the other points of interest on the American side, and after visit the Queen's domains.

Our next point of interest will be the

Whirlpool Rapids.

This wonderful spot is about two miles from the Falls, down the rushing green river which, flowing at profound depth between high banks, looks so quiet yet sullen after leaving the howling abyss at the foot of the falls. At Whirlpool Rapids! what a change! The whole force of the water concentrates itself here, it seems as though it would tear asunder the steep, wooded hills that enclose it, so wild and startling it its terrific power; as far as the eye can reach the water thunders down in seething heaving masses of foam, throwing up streams of water covered with spray, and in places whirling it up into angry billows twenty or thirty feet above the head of the spectator standing on the shore. It is deafening in its roar, and here, even more than at the brink of the Falls, we can have a realization of the terrific force of Niagara. For a this point which is only about 300 feet wide, the united waters of Lake Superior, Michigan, St. Clair, Huron and Erie, thunder along at the rate of 27 miles an hour. According to Sir Charles Lyell's calculations, fully fifteen hundred million cubic feet of water rush through this gorge every minute. Estimated depth 300 feet.

To promote the comfort and convenience of visitors, a double elevator was erected in 1869, at a cost of twenty thousand dollars, and is a perfect specimen of mechanical skill and ingenuity. The machinery is worked by water-power, transmitted from a wheel placed some 300 feet below the top of the bank. Having returned to our carriage, let us proceed one mile further down the river to the

Whirlpool.

The bason containing the Whirlpool is nearly circular and, together with the waters, form a very picturesque scene. But as to the pool itself, it must be acknowledged that many are disappointed with its appearance. It is not, as many suppose, in the shape of a vast caldron or pool with the out-let at the bottom, with the centre depressed, but on the contrary the water is several feet higher in the centre than at the sides. The pool is formed by the peut-up action of the water, and in its bewildered course to find an outlet, is forced around and around the basin. We cannot illustrate it more plainly than to compare the river to a ferocious animal who has never known defeat, that has suddenly, by his own carelessness, fallen into a pit fall. His first impulse is to rush around the outer edge of the pit, in frantic but futile efforts to escape. This passage, when found by the river, appears to be choked and wholly inadequate to carry off the vast amount of water, yet it has answered every purpose for thousands of years.

The grounds around the Whirlpool, belong to the Deveaux College (a school for orphan boys). The proceeds from the admission fees, go to the fund for its support. Taking the whole view of the Whirlpool, College, and drive, the average visitor will be pleased with the trip.

We have now visited the principal points on the American side; we will take our carriage again and, leaving the United States, proceed to Canada. As most tourists wish to go by one bridge and return by the other, we will first cross the old or

Rail-way Suspension Bridge.

This bridge is a noble and stupendous structure; it combines in an eminent degree, strength with elegance of structure. It is owned by a stock company, and cost $ 500,000. Architect Mr. John A. Roebling of Trenton, New-Jersey. This bridge is of enormous strength, and forms communication between the United-States and Canada, over which the cars of the Grand Trunk and Erie Railroad pass withont causing much vibration. It was commenced in 1852; the first cars passed over it on the 8th of March 1855. The road for carriages is suspended 28 feet below the rail-road line. The bridge is now wholly composed of iron and steel except the floor of the carriageway, $ 100,000 having just been expended in taking out the parts made of wood and substituting iron and steel. It is a remarkable fact, that the traffic was not impeded while these extensive repairs were going on. The following statistics in regard to this great structure will not be out of place here:

Length of span from centre to centre of towers 825 feet
Height of towers above the rack, American side 88 »
 » » » » » » Canada side . 78 »
 » » » » » floor of Railway . 60 »
 » » track above the water . . . . . . 258 »
Number of Wire cables . . . . 4 »
Diameter of each cable . . . . 10¼ inches
Number of N° 9 wires in each cable . . . . 3,659 tons
Ultimate aggregate strength of cables . . . . 12,400 »
Weight of superstructure . . . . 800 »
Maximum weight cables and stays will support . 7,309 »

 Grand Trunk through trains east and west run across this bridge, affording passengers a fine view of the whirlpool Rapids where the famous Capt Webb was drowned, the river below, and a distant view of all the Falls.
 The Grand Trunk railway, with its powerful and direct connections, and extensive and continuous through line, is a favorite route. Ever alive to the interests of its patrons, it has yearly improved its track, rolling stock and motive power; it passes through a section of country in which there is a great variety of grand and beautiful scenery, and in all that makes a route desirable to the traveling public this line cannot be surpassed.

In close proximity we perceive the new

Cantilever Bridge.

This novel structure is a double track railway Bridge, built by the Michigan Central railway and connects that railway with the New-York Central. This structure is on a new principle never before illustrated by any large work actually finished. Two similar bridges, however, are now being constructed—one the new **Tay Bridge** over the Firth of Forth, Scotland, and the second for the Canadian Pacific Railway over the Frazer River, British Columbia. Bridges built after the new design are known as cantilever bridges.

The waive motion perceptible on a suspension bridge is not felt on this structure. From the tower foundations up, the whole bridge is made of steel, and is strong enough to bear two of the heaviest freight trains extending the entire length, and under a side pressure of wind at seventy five miles per hour, and even then it will be strained to only one fifth of its ultimate strength.

The following are the dimensions.

Total length of Bridge proper 910 feet.

Length of canti-levers, 375 and 395 feet.

Length of fixed span 125 feet.

Length of clear span across the river, 500 feet.

Height of Abutments, 50 feet.

Height of steel towers, 130 feet.

Height of clear span above the river, 245 feet.

Total weight resting on steel columns, 1,600 tons.

All Michigan Central through trains East and West run across the bridge, affording passengers an excellent bird's eye view of the Falls, the river below, and whirlpool Rapids. The Michigan Central is a popular route of travel between Buffalo, Niagara Falls, Detroit, Toledo and Chicago.

After crossing the bridge the drive on the Canada side, is very fine, as it gives an almost uninterrupted view of the river below the Falls in all of its incessant changes. We may mention, in passing that there are two caves, Catlin's cave and the Giant's cave, between the Bridge and the Falls, on the American side, and Bender's cave on the Canada side. They are, however, not worthy of notice.

The next stopping point is, in front of the

American Falls.

Comprising the American and Centre Fall. These Falls are one quarter of a mile wide, and have a perpendicular height of 164 feet. It is estimated by Sir Charles Lyell that not less than one hundred and fifty millions of cubic feet of water pass over these (2) Falls every minute. They are characterized by an irregularity that gives them a wild and singular beauty. The outline is far-projecting and deeply indented. The water flows over a broad, billowy stream, and is thrown out by craggy points in a hundred places, so that it passes down in a snow-white drapery, and possessing so much beauty and variety that it delights while it awes one almost forgets its immensity while contemplating its singular beauty. Driving on a short distance we come to the famous

Table Rock.

In alluding to this view, Charles Dickens says: — „It was not till I came on Table Rock, and looked on the fall of bright green water, that it came upon me in its full might and majesty. The Niagara was for ever stamped upon my heart, an image of beauty, to remain there, changeless and indelible, until its pulses cease to beat for ever."

Table Rock is no longer the extensive platform that it once was — large portions of it having fallen from time to time. In 1818, a mass of 160 feet long and 40 feet wide broke off and fell into the boiling flood; and in 1828, three immense masses fell, with a shock like an earthquake. Again, in 1829, another fragment fell; and in 1850, a portion of about 200 feet in length and 100 feet thick. On one of these occasions, some forty or fifty persons had been standing on the rock a few minutes before it fell! The work of demolition still goes on, for another portion of Table Rock fell in 1857. In 1867, a large crack or seam having formed around it near the road, it was deemed unsafe, and the Canadian Government caused it to be blasted away; so now all that remains of the once famous Table Rock is a huge mass of rock at the edge of the river below the bank. It overhangs the terrible caldron close to the great

Horse shoe Fall.

The Horse shoe Fall extends from the Canada shore to Goat Island, the width being estimated at 2376 feet; the perpendicular height is 158 feet. It derived its name from its shape; but it has

much altered since it was named. The curve of the Fall has now
little the resemblance of a horse-shoe, having rather the shape of
an inverted letter A.

The volume of water that goes over this Fall is enormous.
It is estimated that the sheet is fully 20 feet thick in the centre
or where it looks so green ; an estimate which was corroborated
in a singular manner in 1829. A ship ramed the Detroit, having
been condemned, was brought and sent over the falls. On board
were put a live bear, a deer, a buffalo, and several smaller ani-
mals. The vessel was almost knocked to pieces in the rapids, but
a large portion of her hull went over entire. She drew eighteen
feet of water, but did not strike the cliff as she took the
awful plunge.

Fully thirteen hundred and fifty millions of cubic feet of water
pass over this Fall every minute. When the sun shines, it adds
much to the magnificent grandeur of this scene ; a beautiful rainbow extending at times from the American to the Horse shoe Fall
can be seen. A spiral stair case, with dressing rooms and guides,
enables the visitor to go down and part way under tho Fall. The
view here is awfully grand. As we look upwards at the frowning
cliff that seems tottering to its fall, and pass under the thick
curtain of water — that it seems as if we could touch it — and
hear the hissing spray, and are stunned by the deafening roar that
issues from the misty vortex at our feet, an indescribable feeling
of awe creeps ower us, and we are again impressed with the
tremendous magnificence of Niagara.

New Suspension Bridge.

This bridge is not as massive in its construction as the lower
one, but more beautiful in architecture. It is designed for foot
passengers and carriages. A favorite American writer speaks of
this bridge as follows; "Over the river, so still with its oily eddies

and delicate wreaths of foam, just below the Falls, they have of late years, woven a web of wire, high in air, and hung a bridge from precipice to precipice. Of all the bridges made with hands, it seems the slightest, most etherial. It is ideally graceful, and droops from its slight towers like a garland. It is worthy to command, as it does the whole grandeur of Niagara, and to show the traveller, the vast spectacle, from the beginning of the American Fall to the farthest limit of the Horse shoe. This is the longest Suspension Bridge in the world. Its great length, symetrical form, graceful curve and obvious strength, cannot fail to strike every beholder with equal surprise and pleasure.

It is owned by a stock company, and cost Doll. 250,000. The following particulars in regard to this structure may be of interest.

The span between the centre of towers is . . . 1,268 feet
Height above surface of river . . . . . 190 »
Height of towers above rock on Canada side . . 105 »
» » » » » » American side . . 100 »

Base of towers 28 feet square, and top 4 feet square. The bridge is supported by two cables, composed of seven wire ropes each, which contain respectively 133 number 9 wires. The weight of these wire ropes per lineal foot is 9lbs, and the diameter of the cable is 7 inches. The aggregate breaking strain of the cable is 1,680 tons. The weight of bridge and appurtenances is about 250 tons. On arriving once more on the American shore we have visited the principal points which comprise all that may be called "Seeing the Falls", and travelled about 12 miles.

With the tourist who has the time and inclination we will now, point out several localities in the vicinity, which are worthy of attention, particularly on account of their historical associations We mention first

Lundy's Lane Battle-ground.

This historical spot upon which he bloody battle was fought, on the 25th day of July, 1814, between the British and American forces, is about a mile and a half west of the Falls on the Canada side. The number of killed and wounded on both sides was about equal, and both parties, as a matter of course, claim the victory.

The Devil's Hole.

About three and a half miles from the Falls, on the American side, is a de‚p, gloomy chasm in the bank of the river, about 120 feet deep. Overhanging this dark cavern is a perpendicular precipice, from the top of which falls a small stream called the Bloody Run. This stream obtained its name from the following tragical incident: — During the French war in 1763, a detachment of British soldiers who were conveying provisions in waggons from Fort Schlosser to Fort Niagara, were here surprised by a party of Indians, then in the pay of the French. The savages, who were numerous, scattered themselves along the hill sides, and lay concealed among the bushes until the English came up, and had passed the precipice; then, uttering a terrific yell, they descended like a whirlwind, and before the soldiers had time to form, poured into their confused ranks a withering volley of bullets. The little stream ran red with blood, and the whole party — soldiers, waggons, horses and drivers — were hurled over the cliff into the yawning gulf below, and dashed to pieces on the rocks. Only two escaped to tell the tale: the one a soldier, the other a Mr. Stedman, who dashed his horse through the ranks of his enemies, and escaped amid a shower of bullets.

Three miles further down is

Top of Mountain.

There are no associations of interest connected with it; it is visited simply on account of the pleasant drive and the grand view to be had from its elevated position. Just below is the village of Lewiston; on the opposite shore Queenstown; between them flows the now quiet river, calm and majestic in its recovered quietude. In the far distance, on either side, stretches the richly wooded landscape, dotted with villas and cottages. This is indeed a fine view, and well repays the exertion of the drive. We see opposite, on Queenstown Heights (Canada side).

Brock's Monument.

This monument was erected by the British Government, in memory of Sir Isaac Brock, the general commanding the army at the battle of Queenstown Heights, on the 13th day of October, 1812. His remains, and those of his aid-de-camp, Col. John Mc. Donald, who died of wounds received in the same battle, are buried here. The first monument was completed in 1826, and was blown up in 1840 by a man named Lett, who was afterwards imprisoned for this dastardly act. The present handsome shaft was erected in 1853. The height is 185 feet; the base is 40 feet square by 30 feet high; the shaft is of freestone, fluted, 75 feet high and 30 feet in circumference, surmounted by a Corinthian capital, on which stands a statue of the gallant General. The view from this monument is most impressive. Niagara is spread out before you, the beautiful valley for seven miles to Lake Ontario, in one grand panoramic view, and the lake beyond, studded with white sails, is one which is excelled nowhere in the country.

Should the visitor be in Niagara on Sunday he will find a very interesting service by attending church at the

Indian Village.

They have two churches, or meeting houses, here, in which the services are conducted in the Indian dialect, then translated for the benefit of those who do not understand it. One house is of the Baptist persuasion, the other Presbyterian: for, of course, the ancient superstitions of the race have faded away before the strong light of Christianity, and the Great Spirit is worshipped only in the name by which the white man calls it. From this village comes all the beautiful bead-work, bows and arrows, Canoes, &c., seen in the bazaars at Niagara. It is eight miles from the Fall.

It is proper to mention that at the
Whirlpool Rapids, Canada Side.

The view is similar to the one we have seen on the American side, being just opposite. This point of interest is generally known as the Whirlpool Rapids Park, and comprises the natural uplands of the river bank, which at this point, are 250 feet high, as well as a road at the base of the cliff, which follows the course of the river, and has been excavated from the rock. In the warm days of summer this is a most delightfully cool and shady retreat, the cliff forming a natural protection from the rays of the sun, while the immediate presence of the swift-rolling waters ensures a perennial coolness. Two means of access to the water's edge are provided, the first being a series of steps forming a long flight of stairs, and the other a unique inclined railway operating two cars running by the specific gravity of water in the tanks under each car filled from a spring at the top of the cliff and emptied on the arrival of the cars at the foot of the incline. The ascent or descent is made in 1½ minutes, the loaded car from above being the motive power used to hoist the lighter car from below.

An other inclined railway enables us to reach the water's edge at the
Whirlpool Canada Side.

The view does not materially differ from the one previously described on the American side, and is much visited.

We think it right to say that the illustrations with which this little book is embellished may be depended on as being minutely correct, having been copied from photographs, made by Niagara's distinguished artists, whose beautiful views of Niagara scenery are so well known to the public. In the preceding pages we have visited the various points around the Falls and pointed out the places of interest in their neighbourhood. We will now, in a few words, give the reader a brief history of some of the local surroundings and events.

Museum

is situated on Canal St., opposite Prospect Park. — The building was erected expressly for the purpose, is a large and massive stone edifice.

The collection is immense having been gathered at great expense from all parts of the world. — The collection of Egyptian Mummies is said to be the best in America. — It is well worth the ad-*mission fee Cyclorama of the Battle of Gettysburg* is also on Canal St. and is much visited by tourists.

"Maid of the Mist".

This little steamer used to run from her dock near the Railway Suspension Bridge up to the Falls, passing so close to the falling waters as to seem to those looking down upon her, to almost pass under the sheet. Water-proof garments were provided, and the trip was an exciting one. The steamer was built expressly for this brief voyage, being an excellent boat of 170 tons burden, with an engine of above 100 horse-power. The owners having found her unprofitable, she was sold to a Montreal firm, with the condition that she should be safely navigated through the Rapids and Whirlpool. This dangerous experiment was undertaken by Joel R. Robinson, (the hero of Niagara) with but two assistants — Mc. Intyre and Jones. She left her moorings, near the Bridge, June 15th, 1861, and swung boldly out into the river, to try one of the most perilous voyages ever made. She shot forward like an arrow of light, and with the velocity of lightning passed on, as many supposed, to meet her doom. Many beheld this hazardous, daring adventure, expecting every instant she would be dashed to pieces and disappear for ever. Amazement thrilled every heart, and it appeared as if no power short of Omnipotence could save her. "There! there"! was the suppressed exclamation that escaped the lips of all. "She is lost! She is lost"! But, guided by an eye that dimmed not, and a hand that never trembled, she was piloted through those maddened waters by the intrepid Robinson in perfect safety. The boat lost her funnel, but otherwise, received no injury. Robinson had performed many hazardous exploits in saving the lives of persons who had fallen into the river, yet this last act, in taking the "Maid of the Mist" through the Rapids and Whirlpool, is the climax of his adventures. It was a wonderful feat of navigation, and created intense excitement for miles around.

Avery on the Log.

In July, 1853, two men took a boat, and set out for a pleasure sail on the river above the Falls. Nothing was heard of them until next morning, when one of them, named Joseph Avery, was

seen clinging to a log sticking on a rock in the midst of the
Rapids below Goat Island Bridge, between Bath Island and the
main-land. Thousands of people assembled to render the poor
man assistance, and during the day various attempts were made
to rescue him from his perilous position, but without success. At
length a boat was lowered down the Rapids towards the log
to which he clung. It reached the spot, but the rope became
entangled under the log, rendering it useless. A raft was then
let down, and he succeeded in getting on it; but those ropes also
became entangled, and the raft could not be brought to shore.
Another boat was let down to him, but as it reached the raft, it
struck with such force that Avery, who was standing erect, fell
off backward, and in another moment he was swept over the Falls.
His body was never found.

Francis Abbot.

The Hermit of the Falls.

In the month of June 1829, a tall, gentlemanly, but haggard-
looking young man, made his appearance at Niagara Falls. He
brought with him a large port-folio, and several books and musical
instruments. For a few weeks he paid daily and nightly visits to
the most interesting points of Niagara, and at length became so
fascinated with the beauty of the scene, that he resolved to take
up his abode there altogether! No one knew whence the stranger
came. Those who conversed with him asserted that he was
talented, and engaging in his manners and address; but he was
not communicative, and shunned the company of man. At the
end of a few weeks he applied for permission to build for himself
a cottage on one of the Three Sister Islands; but circumstances
preventing this, he took up his residence in an old cottage on
Goat Island. Here the young hermit spent his days and nights
in solitary contemplation of the great cataract: and when winter
came, the dwellers on the mainland saw the twinkle of his wood
fire, and listened wonderingly to the sweet music that floated over
the troubled waters and mingled with the thunder of the Fall.

This wonderful recluse seemed never to rest. At all hours of day and night he might be seen wandering around the object of his adoration. Not content with gazing at the Rapids, he regularly bathed in the turbulent waters; and the bathing place of Francis Abbot is still pointed out to visitors. One day in June 1831, he went to bathe in the river below the Falls. Not long afterwards, his clothes were found still lying on the bank, but Francis Abbot was gone. The waters which he had so recklessly dared, had claimed him as their own at last. His body was found ten days afterwards, at the mouth of the river, whence it was conveyed to the burying-ground, near the thundering Falls he loved so well.

Blondin.

It is frequently asked by visitors, where the intrepid Blondin crossed the river on his rope. In 1859 his rope was stretched from bank to bank, about a mile below the Falls; the length of the rope at this place was about 1,200 feet. In 1860 he removed his rope to a point just below the Railway Suspension Bridge; the width here was 900 feet. He crossed the rope many times, carrying a man on his back, and doing many other daring feats. His last performance was given before H. R. H. the Prince of Wales and Suite, and in the presence of a vast multitude of spectators, who had been attracted to witness the miraculous performance of the wonderful Blondin, many coming over two hundred miles to enjoy the novel treat afforded them. On this occasion Blondin put the climax on all his other achievements by crossing the rope on stilts. The saying, what man can do, so can another, has been fully illustrated. Since that time, many persons have succeeded in crossing the river on a rope and performing similar feats. In 1876, a lady named Signorina Maria Spelterina stretched her rope across the river at this point, and astonished the multitude, by her daring and grace, even rivaling the great Blondin in his most wonderful feats.

Niagara by Moonlight.

It were vain to attempt a description of this magical scene.

Every one knows the peculiar softness and the sweet influence of moonlight shed over a lovely scene. Let not the traveller fail to visit Goat Island when the moon shines high and clear, and view Niagara by her pale, mysterious light.

Niagara in Winter.

In all it phases this wondrous cataract is sublime, but in winter, when its darkgreen waters contrast with the pure white snow, and its frosty vapour spouts up into the chill atmosphere from a perfect chaos of ice and foam, there is a perfection of savage grandeur about it which cannot be realized in the green months of summer. At this season. Ice is the ruling genius of the spot. The spray which bursts from the thundering cataract encrusts every object with a coat of purest dazzling white. The trees bend gracefully under its weight as if in silent homage to the Spirit of the Falls. Every twig is covered, every bough is laden; and those parts of the rocks and trees on which the delicate frost-work will not lie, stand out in bold contrast. The trees on Goat Island and in Prospect park seem partially buried; the bushes around have almost disappeared; the buildings seem to sink under their ponderous coverings of white; every rail is edged with it, every point and pinacle is capped with it; and the cold, dismal-looking water hurries its green flood over the brink, and roars hoarsely as it rushes into the vortex of dazzling white below. When the sun shines, all becomes radiant with glittering gems; and the mind is almost overwhelmed with the combined effects of excessive brilliancy and excessive grandeur.

During the winter immense masses of ice descend the river, pass over the Falls, and form an ice bridge below. This ice bridge generally extends from the Horse-shoe Fall, to about a quarter of a mile above the Rail-way Suspension Bridge, and is crossed by hundreds of foot passengers during the winter. When the river rises, the slender hold that binds it to each shore is broken, and the bridge disappears, sometimes in one night, at other times it lasts until May.

The roar of the Falls.

This much depends upon the wind, and the state of the atmosphere. Sometimes every door and window, the least ajar, for a mile around, will tremble, and the roar may be heard from fifteen to twenty five miles. At other times it cannot be heard half a mile away. In a few instances the roar has been heard at Toronto, a distance of forty-four miles.

The river above the Falls.

The descent of the Niagara River — which, let it be borne in mind, flows northward — is only about fifteen feet in the first fifteen miles from Lake Erie, and the country around is almost on a level with the river's banks. At this part the Niagara varies from one to three miles in Width, has a tranquil current, and is lake-like in appearance, being interspersed with low, wooded islands. At the head of the rapids it assumes a totally different appearance, and descends about sixty feet in a mile, over an uneven bed of limestone, and, after being divided into two sheets by Goat Island, plunges down about 164 feet perpendicular at the Falls. Above the rapids islands are numerous. Indeed the river is studded with them, from Lake Erie all the way down to the Falls. There are 37 of them, if we may be permitted to count those that are little more than large rocks. Grand Island is the largest, being 12 miles long and 7 broad. It divides the stream into two branches. Navy Island is just below it. Here the French built their ships of war in 1759. This island was the resort of the rebel leaders in 1837. It has an area of 304 acres. Our space forbids further notice of these islands.

The River below the Falls.

Just below the Falls the river narrows abruptly, and flows rapidly through a deep gorge, varying from 200 to 400 yards wide, and from 200 to 300 feet deep. This gorge, or chasm, extends from the Falls to Queenstown, a distance of seven miles; in the

course of which the river descends 100 feet, and then emerges
on the low, level land lying between Queenstown and Lake Ontario
— a further distance of seven miles. The descent here is only
about four feet altogether, and the flow of the river is placid.
The chasm is winding in form, and about the centre of its course,
makes a turn nearly at right angles, forming the well-known
Whirlpool.

Geology of Niagara.

The geological features of the district around Niagara are
very remarkable, and the Falls afford a fine example of the power
of water to form an excavation of great depth and considerable
length in the solid rock. The country over which the river flows
is a flat table-land, elevated about 330 feet above Lake Ontario.
Lake Erie, situated in a depression of this platform, is about 36
miles distant from Ontario, lying to the south-west. This table-
land extends towards Queenston, where it terminates suddenly in
an abrupt line of cliff, or escarpment, facing towards the north.
The land then continues on a lower level to Lake Ontario

Such are the various appearances and peculiarities presented
by the River and Falls of Niagara, the causes of which we shall
endeavour to explain.

The escarpment at Queenston, and the sides of the great ravine,
have enabled us in the most satisfactory manner to ascertain the
geological formations of the district, and to account for the present
position of the Falls, as well as to form, on good grounds, an
opinion as to the probable working of this mighty cataract in the
future. It has long been supposed that the Falls originally plunged
over the cliff at Queenston, and that they have gradually eaten
their way back, a distance of seven miles, to their present position.
It is further conjectured that they will continue to cut their way
back, in the course of ages, to Lake Erie, and that an extensive
inundation will be caused by the waters of the lake thus set free.
Recent investigation has shown, however, that this result is highly
improbable, we may almost say impossible ; that the peculiar quality
and position of the strata over which the river flows are such,

that the Falls will be diminished in height as they recede, and their recession be altogether checked at a certain point.

It has been ascertained beyond all doubt that the Falls do recede ; but the rate of this retrograde movement is very uncertain; and, indeed, we have every reason to believe that the rate of recession must of necessity in time past have been *irregular*. The cause of this irregularity becomes apparent on considering the formations presented to view at the escarpment and in the chasm. Here we find that the strata are nearly horizontal, as indeed they are throughout the whole region, having a very slight dip towards the south of twenty-five feet in a mile. They all consist of different members of the Silurian series, and vary considerably in thickness and density. In consequence of the slight dip in the strata above referred to, the different groups of rock crop out from beneath each other, and thus appear on the surface in parallel zones or belts; and the Falls, in their retrograde movement, after cutting through one of these zones, would meet with another of a totally different character; having cut through which, a third would succeed, and so on.

In all probability Niagara originally flowed through a shallow valley, similar to the above the Falls, all the way across the table-land to the Queenston Heights or escarpment. On this point Sir C. Lyell writes: "I obtained geological evidence of the former existence of an old river-bed, wich, I have no doubt, indicates the original channel through which the waters once flowed from the Falls to Queenston, at the height of nearly 300 feet above the bottom of the present gorge. The geological monuments alluded to consist of patches of sand and gravel forty feet thick, containing fluviatile shells of the genera Unio, Cyclas, Melania, &c., such as now inhabit the waters of the Niagara above the Falls. The identity of these fossil species with the recent is unquestionable, and these fresh-water deposits occur at the edge of the cliffs bounding the ravine, so that they prove the former extension of an elevated shallow valley, four miles below the Falls — a distinct prolongation of that now occupied by the Niagara in the elevated region between Lake Erie and the Falls."

At the escarpment the cataract thundered over a precipice twice the height of the present one, to the lower level. This lower level, as shown by Hall's Report on the Geology of New York, is composed of red shaly sandstone and marl.

The formations incumbent upon this, exhibited on the face of the escarpment, are as follows; 1. Gray quartzose sandstone; 2. Red shaly sandstone, similar to that of the low level, with hin courses of sandstone near the top; 3. Gray mottled sandstone; 4. A thin bed of green shale; 5. Compact gray limestone; 6. A thick stratum of soft argillo-calcareous shale, similar to that which now lies at the base of the Falls; 7. A thick stratum of limestone, compact and geodiferous, similar to the limestone rock which forms the upper part of the Falls. This is all that we have presented to us in the escarpment; but we may observe, parenthetically, that if we proceed backwards towards Lake Erie, we cross the zone of limestone, and at the Falls discover another stratum of thin-bedded limestone overlapping it, in consequence of the southerly dip before referred to. Further back still we find the Onondaga salt group, which extends, superficially, almost to Lake Erie, where another limestone formation appears.

Now, had there been no dip in the strata of the tableland between Lake Erie and Queenston, it is probable that the Falls would have continued to recede *regularly*, having always the same formations to cut through, and the same foundation to fall upon and excavate. But in consequence of the gentle inclination of the strata to the south, the surface presented to the action of the Falls has continually varied, and the process of recession has been as follows:—

First, the river, rolling over the upper formation of hard limestone, to the escarpment, thundered down a height about double that of the present Falls, and struck upon the red shaly sandstone of the plain below. This being soft, was rapidly worn away by the action of the water and spray, while the more compact rocks above, comparatively unaffected, projected over the caldron, and at length fell in masses from time to time as the undermining process went on. But as the Falls receded, the belt of red sandstone was gradually crossed, and the gray quartzose sandstone became the foundation of the group, and the recipient of Niagara's tremendous blows. This rock is extremely hard; here, therefore, the retrograde movement was probably retarded for ages; and here, just at the point where the Falls intersected this thin stratum of quartzose sandstone, the whirlpool is now situated.

The next formation on which the Falls operated was the red shaly sandstone, similar to the first; which, being soft, accelerated the recession. This went on at increased speed until the stratum was cut through, and the third formation was reached. Here again an alteration in speed occurred as before. The last that has been cut through is the fifth stratum, compact gray limestone, on which the cataract now falls.

The formation now reached, and that on which Niagara is operating at the present day, is the soft argillo-calcareous shale. It extends from the bottom of the precipice over which the water plunges, to nearly half-way up, and is about eighty feet thick. Above it lies the compact refractory limestone, which forms the upper formation at this point. This also is about eighty feet thick; and here we see the process of excavation progressing rapidly. The lower stratum, being soft, is disintegrated by the violent action of the water and spray, aided in winter by frost; and portions of the incumbent rock, being thus left unsupported, fall down from time to time. The huge masses of undermined limestone that fell in the years 1818 and 1828, shook the country, it is said, like an earthquake.

This process is continually altering the appearance of the Falls. Sir Charles Lyell, in his geological treatise on this region, says: "According to the statement of our guide in 1841 (Samuel Hooker), an indentation of about forty feet has been produced in the middle of the ledge of limestone at the lesser Fall since the year 1815, so that it has begun to assume the shape of a crescent; while within the same period the Horse-Shoe Fall has been altered so as less to deserve its name. Goat Island has lost several acres in area in the last four years; and I have no doubt that this waste neither is, nor has been, a mere temporary accident, since

I found that the same recession was in progress in various other waterfalls which I visited with Mr. Hall in the State of New York."

The rate at which the Falls now recede is a point of dispute. Mr. Bakewell calculated that, in the forty years preceding 1830, Niagara had been going back at the rate of about a yard annually. Sir Charles Lyell, on the other hand, is of opinion that one foot per annum is a much more probable conjecture. As we have already explained, this rapid rate of recession has, in all likelihood, not been uniform, but that in many parts of its course Niagara has remained almost stationary for ages.

That the Falls will ever reach Lake Erie, is rendered extremely improbable from the following facts: Owing to the formation of the land, they are gradually losing in height, and therefore in power, as they retreat. Moreover, we know that, in consequence of the southerly dip of the strata, they will have cut through the bed of soft shale after travelling two miles further back; thus the massive limestone which is now at the top will then be at the bottom of the precipice, while, at the same time, the Falls will be only half their present height. This latter hypothesis has been advanced by Mr. Hall, who, in his survey, has demonstrated that there is a diminution of forty feet in the perpendicular height of the Falls for every mile that they recede southward: and this conclusion is based upon two facts — namely, that the slope of the river-channel, in its course north- ward, is fifteen feet in a mile , and that the dip of the strata in an opposite or southerly direction is about twenty-five feet in a mile.

From this it seems probable that, in the course of between ten and eleven thousand years, the Falls of Niagara, having the thick and hard limestone at their base, and having diminished to half their present height, will be effectually retarded in their retrograde progress, if not previously checked by the fall of large masses of the rock from the cliff above. Should they still recede, however, beyond this point, in the course of future ages they will have to intersect entirely different strata from that over which they now fall, and will so diminished in height as to be almost lost before reaching Lake Erie.

The question as to the origin of the Falls — the manner in which they commenced, and the geological period at which they first came into existence — is one of great interest; but want of space forbids our discussing that question here. We can make but one or two brief remarks in regard to it.

Sir Charles Lyell is of opinion that originally the whole country was beneath the surface of the ocean, at a very remote geological period; that it emerged slowly from the sea, and was again submerged at a comparatively modern period, when shells then inhabiting the ocean belonged almost without exception to species still living in high northern latitudes, and some of them in temperate latitudes. The next great change was the slow and gradual re-emergence of this country.

As soon as the table-land between Laken Erie and Ontario emerged, the river Niagara came into existence; and at the same moment there was a cascade of moderate height at Queens on, which fell directly into the sea. The cataract then commenced its retrograde movement. As the land slowly emerged, and the hard beds were exposed, another Fall would he formed; and then probably a third, when the quartzose sandstone appeared. The recession of the uppermost Fall must have been retarded by the thick limestone bed through which it had to cut: the second Fall, not being exposed to the same hindrance, overtook it; and thus the three ultimately came to be joined in one.

The successive ages that must have rolled on during the evolution of these events are beyond the power of the human intellect to appreciate, and belong to those "deep things" of the great Creator, whose ways are infinitely above our finite comprehension. It is roughly calculated that the Falls must have taken at least 35,000 years to cut their way from the escarpment of Queenston to their present position; yet this period, great though it is in comparison with the years to which the annals of the human race are limited, is as nothing when compared with the previous ages whose extent is indicated by the geological formations in the region around Niagara.

The first white visitor.

The first white man who saw the Falls, as far as we have
any authentic record , was Father Hennepin, a Jesuit missionary,
sent out by the French among the Indians, as early as 1678, over
200 years ago. His descriptions were visionary and greatly exag-
gerated. He described the Falls to be six or seven hundred feet
high, and that four persons could walk abreast under the sheet
of water without any other inconvenience than a slight sprinkling
from the spray. We will not attribute this wild and fanciful description to a want of candor, or intention to deceive. The fact
probably was, he had no means of measuring its height, and undoubtedly got his account from the Indians, which very likely
would be incorrect.

A Daring feat.

In 1879, a man named Peer n....e his appearance, and announced that he would j[·]   from the New Suspension Bridge
into the river. When he to.   ..nat he intended to do, the people
naturally considered him a lunatic, but on the 21 st May he did
make this, the "greatest of all leaps". He had a mechanical
contrivance to keep him from turning in the descent; it is fair to
say, that it did not break the fall. He put a board out from the
bridge and stood looking at the multitude who had been attracted
to the place with the expectation of seeing him jump and get
killed. However, he performed this wonderful feat, and received
only a few slight bruises. He stepped off the board; went down
feet foremost 190 feet into the river, striking the water with a
report like a gun. Time of descent four seconds. He proposed to
repeat this feat on the fourth of July, but the inducements not
being sufficient. and being somewhat afraid, he abondoned the
foolhardy undertaking.

Capt. Webb's Fatal Swim.

Capt. Matthew Webb, the famous English swimmer, made
the attempt to swim through the Rapids and Whirlpool of Niagara

River on the afternoon of July 24th, 1883, and lost his life in the effort. As he had publicly announced he would do, Capt. Webb left the Clifton House on the Canada side, at 4 o'clock, New York time, and proceeded down the bank to the Ferry landing. Here he stepped into a small boat manned by J. McCloy, ferryman, and was rowed down the river to opposite the old Pleasure Grounds, just above the Maid of the Mist landing, when at 4.25 he jumped from the boat into the river and swam leisurely down to the Rapids which were to engulf him. At 4.33 he passed under the Railway Suspension Bridge into the Rapids. At 4.35 he reached the last of the Rapids before entering the mouth of the Whirlpool. Here he was seen to sink below the crest of the Rapids and he never appeared on the top of the water again. Some of the spectators think that they saw his body near the top of the water 50 or 100 feet below the spot where he disappeared from the surface, but all agree that he never again came to the top of the water. When he passed under the Suspension Bridge he seemed to have perfect control of himself and this all accounts agree he maintained until he reached the height of the Rapids opposite the Whirlpool Rapids Elevators. At this point accounts differ as to his appearance. Some say that he maintained his equipoise through that terrible channel, while others say that he appeared like a drowning man, sport of the waves. But certain it is that after passing the fiercest of these Rapids he momentarily regained control of himself, for the spectators on the Whirlpool Grounds on both sides of the river saw him as he emerged into the comparatively still waters that intervene before the Whirlpool is reached, rise upon the surface and throw at least a third of his body above the angry waters. Then he seemed to swim on top of the water for a hundred feet, when he disappeared forever. The spectators who saw him disappear waited a few seconds to see him reappear on the crest of the current, but they watched in vain. Then all rushed to the water's edge in the hope he might have passed safely through the great maelstrom and have landed somewhere on the bank. The banks were thoroughly searched, but no trace of him was discovered, and a belief began to grow that he would never be seen alive. The only hope left was that he might have passed unseen into the lower rapids and going through them had left the river at Lewiston or Queenston, but as the hours passed this hope was dissipated and the fact was made known that the brave and intrepid Webb had met more than his match in Niagara's mad waters.

Finding of the Body.

Nothwithstanding the fact that Webb's fatal swim was witnessed by a large number of people, much doubt was expressed as to whether he had actually made the attempt, or if he had, that he might have left the river alive at some point beyond the observation of the spectators. All uncertainty on these points were, however, removed by the finding of Capt. Weeb's body about noon on Saturday, July 28th. 1883, a little more than four days after his disappearance.

The body was found by Richard W. Turner, of Youngstown, about a mile and a half below, Lewiston, floating in Niagara river, about noon on Saturday. He tied the body to the shore and rowed over to Lewiston for help, and the corpse was finally towed to a boat house there. All this took considerable time and the body was not fully identified till about 3 o'clock in the afternoon, when it was recognized by Felix Nassioy, clerk of the New York Central House, and Charles Wiedenman, of Suspension Bridge, who were the last persons to speak to Captain Webb before he entered the Rapids, ailing him from a small boat in which they were crossing the river. The bodies of the two Indians who were drowned the day before Captain Webb made the attempt, were also recovered during the day, and this gave rise to some confusion at first but the investigation which followed removed all doubt.

Capt. Webb's Brief History.

Capt. Matthew Webb was a native of England and 35 years of age. His father lives in Shropshire England, and there were 13 children in the family, eight being boys. He learned to swim when eight years old, being encouraged in his ventures by his father. While yet a mere youth he ran away to sea, and during his career before the mast became famous for his swimming feats, several of which were performed in saving human life. In 1872, while in South Africa, he won his first laurels as a public swimmer, and in the year following received a purse of $ 500 from the passengers of the steamer Russia and a medal from the humane society of London for saving the life of a sailor who was washed overboard. The achievement that gave him international fame was swimmiug the English chanuel naked and without aid of any kind, on which occasion he was in the water from 1 p. m. to 11 a. m. the next day. When he was dragged out of the water at the close of this exploit he was presented with $ 25,000 by the Prince of Wales. On one occasion he swam from Sandy Hook to Manhattan Beach during a storm that drove vessels into the harbor. In July, 1882, he beat Wade at Coney Island for the American championship and at different times has performed wonderful feats in the water, of which no record has been made. Recently he has made his home in Boston, where his wife, also of English birth and but a few years resident in America, and two children mourn the loss of his untimely death. Never were physical prowess and courage worse applied than in the brave fellow's last adventure, which, even if successful, would have been of no pratical service to the world.

The Hotels.

The hotels are excellent, well kept, and compare favorably with the hotels of any other locality in America. Considering the elegance of the accommodations, the quality and sumptuousness of the fare provided, the charges are very moderate.

The chief of them are International Hotel, Cataract-House, Spencer House, Prospect House and Hotel Kaltenbach, on the American side, and on the Canada side Clifton House.

The International.

This magnificent hotel is situated on the corner of Main and Falls Streets, and has a frontage of 650 feet. Its appointments are first class in every respect, with accommodations for about 500 guests. Under the present management its popularity is greater than ever. The proprietors, Messrs. Cluck, Ware and Delano, are thorough business men, and know how to keep a first class hotel.

Cataract House.

This splendid hotel is situated on Main Street, and is too well known to need much comment. It has accommodations for about 500 guests and has all the modern improvements. Messrs. Whitney, Jerauld & Co., are veteran hotel keepers and know how to anticipate the repuirements of their guests.

Spencer House.

This attractive Hotel is situated directly opposite the N. Y. C. depot, and although not so large as the International or Cataract, it shares with them the best public patronage. Unlike them, however, it is open tho year round. Its central location, and its convenient situation with reference to the depot, combine to make it equally agreeable to the tourist or man of business. The proprietor, A. Cluck, has won a wide reputation.

Hotel Kaltenbach.

This fine, new hotel is nicely situated on Buffalo street, facing the rapids; and has already a wide reputation in the hands of Mr. Kaltenbach, whose natural qualifications, together with years of experience, make him one of the best landlords in the village. This hotel is a great favorite with the German tourist, and is also open the year round.

Clifton House.

This commodious hotel is situated near the banks of the river on the Canada side! its balconies command a fine view of all the Falls; it has accommodations for about 300 guests. The proprietors Mr. Colburn have made it a model for luxury, neatness, order, and thorough good Management. Open only during the summer months.

Prospect House

is situated on Union St., it is new, and elegant in all its appointments. This house is open the year round D. Isaacs, Proprietor.

There are many other good hotels on the American side, namely: Niagara House, Falls Hotel, Western Hotel, &c. &c.

Carriages and Hackmen.

Carriages are not scarce, or difficult to find; they can be had at all the hotels, public hack-stands, and almost at every turn. The rates of fare are placed in every carriage in a conspicuous place. When engaged by the hour, the price is, $ 1.50 per hour. When engaged for a number of hours, lower rates generally prevail.

The Hackmen

These innocent, and much abused individuals are noted for their perseverance, the zeal they manifest in their business, their importuning propensities, and their ignorance of the meaning of that little word, "No". Nothwithstandig this, they can be depended on to fulfil any contract they make, and are generally accommodating and honorable in their dealing. In most cases, where complaints of extortion have been made against these hackmen the fault has been with the person making the complaint. In many cases, they expect to deduct from the cariage fare. the amount paid for admission fees at the various points. Visitors should make a bargain with the hackmen before entering the vehicle — in accordance with his rates of fare — and request no gratuitous "extras" at his hands, and there will be no cause for complaint.

Distances.

From Principal Hotels.

Around Goat Island	1½ miles
» Prospect Park	½ »
To New Suspension Bridge	¼ »
» Railway Suspension Bridge	2 »
» Whirlpool Rapids	2½ »
» Whirlpool	3 »
» Devil's Hole	3½ »
» Top of Mountain	6½ »
» Indian Village (Council House) . . .	8 »
» Table Rock, via New Suspension Bridge, or Ferry	1¼ »
» » » » Railway Suspension Bridge . .	4¾ »
» Lundy's Lane Battle Ground	2 »
» Brocks's Monument, Queenstown Heights . .	7 »

Admission Fees and Tolls.

To Goat Island	Free
» Prospect Parks	»
» Cave of Winds (with guide and dress) . .	one dollar
» Inclined Railway (Prospect Park) . . .	10 cents
» Ferry to Canada and return	50 »

To Behind Horse shoe Fall with guide and dress
Canada side . . . . . . .50 cents
» Museum American side . . . . . .50 »
» Lundy's Lane Battle-ground . . . . .50 »
» Whirlpool Rapids (either side) . . . .50 »
» New Suspension Bridge, each way . . . .25 »
» » » » extra for two horse carriage 50 »
» Railway Suspension Bridge, over and return .25 »
» » » » extra for two horse carriage 50 »
» Cyclorama Battle of Gettysburg . . . .35 »

The number of victims whose carelessness or folly has sent over the Falls is quite formidable, and doubtless quite independent of the Indian tradition that the cataract demands a yearly sacrifice of two victims, since no such tradition can be authenticated. We give below a record of some of the recent and most memorable.

Accidents, Suicides and narrow Ecapes.

Sept., 1859, two men were observed clinging to the bottom of a boat about a mile above the Falls, they went down on the outside of the Third Sister Island, and thence over the cataract. One of the unfortunate men was named Johnson, who had some years before been rescued from a perilous position in the Rapids, by Joel R. Robinson.

Sept., 1859, a man calling himself Shields, proposed to jump from a platform, 90 feet into the river. Before making the leap, he went into the river to ascertain the depth, strength of the current etc., and was never seen after!

May, 1864. Mrs. Bender committed suicide on the Canada side of the Falls, where she resided. She walked into the rapids just above the Horse, shoe Fall, and was instantly swept over; her body war never found. She had been insane for many years.

Oct., 1865, a young man named William Duncan undertook to row across the river to Chippewa but went over the Falls.

Parts of his body were found a few days after near the Ferry below the Falls.

Sep., 1866, two men, named Daniel Coffa and Henry Husted, started to row from Chippewa to the American side, they got into the Rapids and went over the Falls. Bodies not found.

Sep., 1869, a gentleman calling himself Carl Schurz, went to the Horse shoe Fall, (Canada side) walked into the Rapids, and was instantly swept over. His body was found a few days after at the Ferry.

While this body was anchored at the Jerry, (Canada side) another terrible accident happened. A party from Providence, R. I., consisting of Mr. Tillinghast and wife, Mrs Fisher, Miss Smith and Miss Balou were in a carriage going down the hill leading to the ferry landing, intending to cross in the ferry boat to the American side when part of the harness gave way which caused the horses to jump and throw the carriage down the precipice. Mr. Tillinghast and the driver sprang from the carriage in time to save themselves, but the ladies were carried over, and fell a distance of forty or fifty feet. Miss Smith was found to have been killed instantly by a blow on the head. Mrs. Tillinghast was bruised about the head and otherwise injured. Mrs Fisher had one wrist fractured and suffered contusions in various parts of the body. Miss Ballou was taken up for dead, but finally recovered consciousness she had three ribs broken, three breaks and one compound fracture of the right arm, cut and bruised about the head, and spine injured. She is still living but is a sufferer from the effects of the fall Dec., 1869, Mr. James Pierce, an old resident to Niagara-Falls, committed suicide by jumping off the Railway Suspension Bridge. His body was never found.

Nov., 1870, Mrs. Margaret Avery, a resident of Chicago, committed suicide by jumping off Goat Island Bridge into the Rapids, and was almost immediately whirled over the American Fall. It was afterwards ascertained that the unfortunate lady was insane. Body not found.

May, 1871, Three young men whose names were not known, attempted to cross the river above the Falls; not being familiar with the currents, they were soon drawn in the rapids, and over the Horse shoe Fall. None of the bodies recovered.

July, 1872, Two men, names unknown, in crossing above the Falls, were drawn into the current and went over.

July, 1863, a young man accompanied by a young lady and boy hired a boat and started for a sail; they were soon in the rapids and over the cataract. The remains of the young lady were found two days afterward, at Youngstown. There was a romantic as well as a tragic feature in the above. It appeared that the young man and young lady had run away for the purpose of getting married; the latter insisting, as a matter of prudence, that her brother should be of the party. The marriage was to have taken place on the day of the catastrophe. The father and the mother of the girl, who had come in pursuit of the runaway, arrived but a short time after!

Sept., 1874, a stranger reached the Falls on the Canada side, was driven to Table Rock, and after viewing the Falls for a time, he asked the carriage driver if he thought "it would kill a man to go over". The driver assured him it would be certain death. Requesting him to mail a letter for him, and without saying anything more, walked into the water, and in a moment was carried over the Falls. His body was never found, nor did his name afterwards become known.

August, 1875. A sad fate befell two estimable young people, residents of the village of the Falls. Miss Lottie Philpott, with two brothers, a sister-in-law, and Mr. Ethelbert Parsons, went through the Cave of the Winds, and climbed over the rocks towards the American Fall, to bathe in the lighter currents that sweep between and over the massive rocks below. With a rash, venturesome spirit, Miss Philpott chose one of the most dangerous currents in which to bathe; she soon lost her footing and fell, Mr. Parsons grasped for her, but failing to catch her, he sprang into the current and both were carried down the stream. Desperately the brave man labored to save her, but of no avail, for the current carried them further from the shore. The horror-stricken witnesses, unable to render any assistance saw, them sink below the surface. Where they disappeared a cloud of mist hid the scene of disaster for a moment, and when their friends next looked for their loved ones the angry waters gave no sign of the tragedy just enacted. Both bodies were subsequenlty recovered at the Whirlpool.

July, 1877. A fatal casualty occurred on the river above the Falls, by which two men lost their lives and a third narrowly escaped meeting the same fate. Charles A. Pierce, Wallace Belinger. and William Flay, all residents of Niagara Falls were sailing on the river, when their boat capsized; they all succeeded in getting hold of the boat and attempted to get it right side up, but after making several futile endeavors, and becoming exhausted by repeated immersions, they gave up the attempt. Pierce and Bellinger then tried to swim for the shore, but having on their clothes they did not go far before they went down to their death. Flay managed to get upon the keel of the boat, from which he was washed several times, but managed to get back again each time in safety. He was rescued from this awful position in an exhausted state, by three young men named Walker. The body of Bellinger was found on Grass Island; the body of Pierce went over the Fall and was afterward recovered in the Whirlpool.

April, 1878. Two young men, Brothers and resi dents of Chippewa while crossing the river above the Falls to their home, got into the rapids and were carried over the Falls. The bodies were after wards recovered.

May, 1879. A well known citizen of Niagara Falls, named Pipus Walker, rowed out into the river, and was soon in the rapids and over the American Fall. He was a good boat-man and well acquainted with the river in every respect, and had he been sober at the time, it is likely this notice would not have been here. His body was recovered.

June, 1879. A lady and gentleman, named Roland, from Belgium, who were on a trip arround the world, visited Niagara Falls, and went over on the Three Sister Islands. Mr. Roland came back alone, naturally very much excited, stating that his wife while stooping to get a drink of water, accidently fell into the rapids and had gone over the Horse-shoe Fall. Suspicions of foul-play were entertained, but there being no proof, Mr. Roland was allowed to go on his way. Before leaving he left a sum of money to defray the funeral expenses in case the body should be found. A few days afterwards the body was recovered near the Falls, and conveyed thence to the burying-ground.

Sep., 1880. A gentleman from Utica, named Knapp, committed suicide under the following circumstance. He arrived at the Falls in the evening, and went to the telegraph office and enquired for a dispatch, there not being one for him, he seemed very much disappointed; he then walked down to the Rapids at the end of Tugby Mammoth Bazaar, where he stepped upon a bench, shot himself in the head, and fell head-long into the rapids; in a few seconds he was over the American Fall. Before committing this rash deed, he took off his coat and vest and laid them on the walk, and attached a paper stating where they were to be sent. He also left his watch and chain, a small amount of money, gold sleeve buttons, kni`:, and several other articles. His body. was recovered soon afte.' in the Whirlpool.

Inside of five years, several men have committed suicide, by jumping from the New Suspension Bridge.

Jan., 1883. Thomas Hilson of Philadelphia committed suicide by jumping into the rapids from Luna Island »Before the war he carried on a prosperous wool trade, and afterwards formed a partnership with Geo. W. Bond & Co., of Boston, the fir: 's name in Philadelphia being Thomas Hilson & Co. About ten years ago the firm became involved in financial difficulty, brought on, so Hilson claimed, by speculative ventures of some of the junior partners. Hilson gathered together all the money he could get hold of, amounting to $ 40,000 or $ 50,000, and decamped. The senior member of the firm, Mr. Bond, happened to be in Europe at the time, and on the arrival of the steamship had Hilson arrested. Mr. Bond succeeded in getting $ 28,000 from the fugitive. This money was paid to the firm's creditors, and the affair was thus settled. In 1876 Gregg Brothers, who had business transactions with Hilson, after his return from Europe charged him with the embezzlement of $ 6,000, and he was arrested while on a visit to Philadelphia during the centennial. He was subsequently tried, but the jury disagreed and the matter was compromised. Hilson's friends are satisfied that he killed himself because he had deliberately come to the conclusion that he could no longer live in the extravagant manner in which he delighted, and, rather than be compelled to exist without »high living,« he preferred to end his life.

»April 9th 1834. Two well known citizens of Suspension Bridge, Thomas Vedder and Van R. Pearson, started out for a ride, they not returning in proper season, a search was instituted. About two o'clock the next morning, the horse and carriage was found on Goat-Island, near the stairs leading to Luna-Island and shortly after the dead body of Pearson was found on Luna-Island, with two bullet holes in his head. Near by lay a portion of the clothing belonging to Thomas Vedder, but no traces could be found of Mr. Vedder. On Thursday June 5th while the bridges were being put in the Cave of the Winds, some of the workmen found the body of Mr. Vedder. There is still some mystery about the sad affair. What passed between the two men as they stood together on Luna-Island that cold evening, will never be revealed until the end of time, when all things are made known.

The preceding is but a partial lest of the known victims. The number of those who have taken the fatal plunge at night, unseen, can never be ascertained.

Many state that while looking into the chasm, an almost irresistible impulse besets them to leap into the fearful flood. We cannot explain this, and why such a feeling should possess the mind is beyond our comprehension, though certain it is that such a feeling does exist in the minds of many.

The following narrow escapes may now be mentioned. In 1839, a man named Chapin, who was engaged in repairing Goat Island Bridge, fell into the Rapids, fortunately the current carried him to the first of two small Islets below. He was rescued from his perilous position by J. H. Robinson, who had more than once bravely rescued fellow-creatures from this dangerous river; and the island was named after him — Chapin Island.

In 1852, a gentleman from Troy, N. Y., while passing over Terrapin Tower Bridge, fell into the river, and was instantly carried to the verge of the precipice and lodged between two rocks. Mr. Isaac Davy assisted by a visitor, rescued him, by throwing lines to him : he had just sufficient strength left to fasten them around his body; then they drew him to the bridge in an exhausted condition.

In 1874, Mr. Wm. Mc Collough, on old resident of the Falls, while engaged in painting the bridge leading to the second Sister Island, accidently fell into the Rapids. When about a quarter of the way down to the spot where he would take the fatal plunge, the current threw the nearly insensible man over a low ledge into a small eddy, where he managed to get upon a projecting rock. He was recued by Conroy, the well known guide, who succeeded in getting to him with a line, by the aid of which, both were landed in safety.

In 1875, an unknown man fell over the bank, on the Canada side, a few rods below the New Suspension Bridge. Those by whom the accident was witnessed, repaired to the spot where he fell, expecting to find his mangled remains. Their surprise was great when they found him unhurt with the exception of a few slight bruises ! This seemed almost incredible, when he had fallen eighty feet perpendicular, and then rolled down the bank sixty feet further. It is, perhaps worthy of remark, that he was intoxicated when he fell over, but sober after.

Legend of the white Canoe.

In days of old, long before the deep solitudes of the West were disturbed by white men, it was the custom of the Indian warriors of the forest to assemble at the Great Cataract and offer a human sacrifice to the Spirit of the Falls. The offering consisted of a white canoe full of ripe fruits and blooming flowers ; which was paddled over the terrible cliff by the fairest girl of the tribe who had just arrived at the age of womanhood. It was counted an honour by the tribe to whose lot it fell to make the costly sacrifice ; and even the doomed maiden deemed it a high compliment to be selected to guide the white canoe over the Falls. But in the stoical heart of the red man there are tender feelings which cannot be subdued, and cords which snap if strained too roughly.

The only daughter of a chief of the Seneca Indians was chosen as a sacrificial offering to the Spirit of Niagara. Her mother had been slain by a hostile tribe. Her father was the bravest among the warriors, and his stern brow seldom relaxed save to his blooming child, who was now the only joy to which he clung on earth. When the lot fell on his fair child, no symptom of feeling crossed his countenance. In the pride of Indian endurance he crushed down the feelings that tore his bosom, and no tear trembled in his dark eye as the preparations for the sacrifice went forward. At length the day arrived; it faded into night as the savage festivities and rejoicing proceeded; then the moon arose and silvered tho cloud of mist that rose from out the turmoil of Niagara; and now the white canoe, laden with its precious freight, glided from the bank and swept out into the dread rapid from which escape is hopeless. The young girl calmly steered her tiny bark towards the centre of the stream, while frantic yells and shout arose from the forest. Suddenly *another* white canoe shot forth upon the stream, and, under the powerful impulse of the Seneca chief, flew like an arrow to destruction. It overtook the first; the eyes of father and child met in one last gaze of love, and then they plunged together over the thundering cataract into eternity !

New Maid of the Mist.

A new steamer called "Maid of the Mist", has been built below the Falls to take the place of her famous predecessor of the same name. I would advise all visitors to take a trip on this boat before leaving Niagara.

Fare, for the round trip, 50 cents.

Father Hennepin's Description
of the Falls, published in 1678.

Betwixt the Lake *Ontario* and *Erie*, there is a vast and pro-digious Cadence of Water, which falls down after a surprizing and astonishing manner, insomuch that the Universe does not afford its Parallel. «Tis true, *Italy* and *Suedeland* boast of some such Things; but we may well say they are but sorry patterns, when compar'd to this of which we now speak. At the foot of this horrible Precipice, we meet with the River *Niagara*, which is not above a quarter of a League broad, but is wonderfully deep in some places. It is so rapid above this Descent, that it violently hur ies down the wild Beasts while endeavoring to pass it to feed on the other side, they not being able to withstand the force of its Current, which enevitably casts them headlong above Six hundred foot high.

This wonderful Downfall is compounded of two cross-streams of Water, and two Falls, with an isle sloping along the middle of it. The Waters which fall from this horrible Precipice, do foam and boyl after the most hideous manner imaginable, making an outrageous Noise, more terrible than that of Thunder; for when the Wind blows out of the South, their dismal roaring may be heard more than Fifteen Leagues off.

The River *Niagara* having thrown it self down this incredible Precipice, continues its impetuous course for two Leagues together, to the great Rock above-mention'd, with an inexpressible rapidity: But having pass'd that, its impetuosity relents, gliding along more gently for other two Leagues, till it arrives at the Lake *Ontario* or *Frontenac.*

Any Bark or greater Vessel may pass from the Fort to the foot of this huge Rock above mention'd. This Rock lies to the Westward, and is cut off from the Land by the River *Niagara*, about two Leagues further down than the great Fall, for which two Leagues the People are oblig'd to transport their goods overland; but the way is very good; and the Trees are very few, chiefly Firs and Oaks.

From the great Fall unto this Rock, which is to the West of the River, the two brinks of it are so prodigious high, that it would make one tremble to look steadily upon the Water, rolling along with a rapidity not to be imagin'd. Were it not for this vast Cataract, which interrupts Navigation, they might sail with Barks, or greater Vessels, more than Four hundred and fifty Leagues, crossing the Lake of *Hurons*, and reaching even to the farther end of the Lake *Illinois*, which two Lakes we may easily say are little Seas of fresh Water.«

The village of Niagara Falls.

The village of Niagara Falls takes its name from the Great Cataract, is situated on the Niagara River, about 22 miles from Buffalo, and is accessible by rail from all parts of the United States and Canada. The population is about 3,900. During the war of 1812, this locality was the scene of many startling events, which have passed into history. The climate is in the highest degree healthy and invigorating. The atmosphere being continually acted upon by the rushing waters, is kept pure, refreshing, and salutary, and is blessed with that which companies can neither purchas nor monopolize — cool breezes from the river.

Besides being a fashionable place of resort, it must eventually become a large manufacturing town. The vast water-power, (probably the best in the world) is just beginning to attract the attention of capitalists. Mills are already built, others are in the course of construction, and capitalists are now negotiating far other sites.

There is established an International Park at this place, according to the plan suggested by Lord Dufferin. The necessity of action became more urgent every year with the growth of the surrounding country and development of manufacturing enterprises. The object of this scheme was not to drive away manufacturers, but simply to exlude them from the immediate proximity of the cataract, while giving every opportunity for industrial expansion by the use of the unrivalled waterpower at a point further down the river. This has done away with the high and numerous tolls, preservad the natural beauty of the Falls, and created a reservation of which both countries are equally proud.

The name "Niagara" is a corruption of the Indian word „Onyakara" supposed to be of the Iroquois language. The meaning of the term is "mighty, wonderful. thundering water.

The state of New-York has purchased Goat Island group, Prospect Park, and a strip of land along the river front about ¾ of a mile long and about 200 feet wide, and made the Islands and Prospect Park free to the world.

Descriptive Pieces.

To Niagara
written at the first sights of its Falls.

Hail! Sovereign of the world of Floods! whose majesty and might
First dazzles, then enraptures, then o'erawes the aching sight :
The pomp of Kings and Emperors, in every clime and zone,
Grows dim beneath the splendor of thy glorious watery throne.

No fleets can stop thy progress, no armies bid thee stay,
But onward — onward — onward — thy march still holds its way:
The rising mists rhat veil 'hee as thy heralds go before,
And the music that proclaims thee is the thund'ring cat'ract's roar!

Thy diadem's an emerald, of the clearest, purest hue.
Set round with waves of snow white foam, anu spray of feathery dew;
While tresses of the brightest pearls float o'er thine ample sheet,
And the rainbow lays its gorgeous gems in tribute at thy feet.

Thy reign is from the ancient days, the sceptre from on high,
Thy birth was when the distant stars first lit the glorius sky ;
The sun, the moon, and all the orbs that shine upon thee now,
Beheld the wreath of glory which first bound thine infant brow.

And from that hour to this, in which I gaze upon thy stream,
From age to age — in winter's frost or summer's sultry beam —
By day, by night, without a pause, thy waves with loud acclaim,
In ceaseless sounds have still proclaimed the great Eternal's name.

For whether, on thy forest-banks, the Indian of the wood,
Or, since his day, the red man's foe on his fatherland has stood;
Whoe'er has seen thine incense rise, or heard thy torrents roar,
Must have knelt before the God of all to worship and adore.

Accept then, O Supremely Great! O Infinite! O God!
From this primeval altar, the green and virgin sod,
The humble homage that my soul in gratitude would pay.
To Thee whose shield has guarded me through all my wandering way.

For if the ocean be as nought in the hollow of Thine hand,
And the stars of the bright firuament in Thy balance grains of sand
If Niagara's rolling flood seem great to us who humbly bow,
Oh, Great Creator of the Whole, how passing great art Thou!

But though Thy power is far more vast than finite mind can scan,
Thy mercy is still greater shown to weak, dependent man:
For him thou cloth'st the fertile earth with herbs, and fruit, and seed;
For him the seas, the lakes, the streams, supply his hourly need.

Around, on high, or far, or near, the universal whole
Proclaims Thy glory, as the orbs in their fixed courses roll;
And from Creation's grateful voice the hymn ascends above,
While Heaven re-echoes back to Earth the chorus — "God is love!"

                                        J. S. BUCKINGHAM.

The Falls of Niagara.

The thoughts are strange that crowed into my brain
While I look upward to thee. It would seem
As if God poured thee from His "hollow hand",
And hung His bow upon thine awful front,
And spoke in that loud voice which seemed to him
Who dwelt in Patmos for his Saviour's sake,
"The sound of many waters;" and had bade
Thy flood to chronicle the ages back,
And notch the centuries in the eternal rocks.
Deep calleth unto deep. And what are we,
That hear the question of that voice sublime?
Oh! what are all the notes that ever rung
From War's vain trumpet, by thy thundering side?

Yea, what is all the riot that man makes
In his short life, to thy unceasing roar?
And yet, bold babbler, what art thou to Him
Who drowned a world, and heaped the waters far
Above its loftiest mountains? — a light wave
That breaks and whispers of its Maker's might!

BRAINARD.

Niagara.

Flow on for ever, in thy glorious robe
Of terror and of beauty. Yea, flow on,
Unfathomed and resistless. God hath set
His rainbow on thy forehead, and the cloud
Mantled around thy feet. And He doth give
Thy voice of thunder power to speak of Him
Eternally — bidding the lip of man
Keep silence, and upon thine altar pour
Incense of awe-struck praise.
     Earth fears to lift
The insect trump that tells her trifling joys
Or fleeting triumphs, mid the peal sublime
Of thy tremendous hymn. Proud Ocean shrinks
Back from thy brotherhood, and all his waves
Retire abashed. For he hath need to sleep,
Sometimes, like a spent labourer, calling home
His boisterous billows, from their vexing play,
To a long dreary calm: but thy strong tide
Faints not, nor e'er with falling hearts forgets
Its everlasting lesson, night nor day.
The morning stars, that hailed Creation's birth,
Heard thy hoarse anthem mixing with their song
Jehovah's name; and the dissolving fires,
That wait the mandate of the day of doom
To wreck the Earth, shall find it deep inscribed
Upon thy rocky scroll.

                        Lo! yon birds,
How bold, they venture near, dipping their wing
In all thy mist and foam. Perchance 'tis meet
For them to touch thy garment's hem, or stir
Thy diamond wreath, who sport upon the cloud
Unblamed, or warble at the gate of heaven
Without reproof. But as for us, it seems
Scarce lawful with our erring lips to talk
Familiarly of thee. Methinks, to trace
Thiue awful features with our pencil's point
Were but to press on Sinai.
                        Thou dost speak
Alone of God, who poured thee as a drop
From His right hand — bidding the soul that looks
Upon thy fearful majesty be still,
Be humbly wrapped in its own nothingness,
And lose itself in Him.
                                        MRS. SIGOURNEY.

We have now done with Niagara and its neighborhood, but
there is a fascination about this mighty Cataract which seems to
chain us to the spot, and when we seek to leave it, draws us
irresistibly back again: Even in describing it, however inadequately
the task may be accomplished, we are loth to lay down the pen
and tear ourselves away. It is a scene which poets and authors
have tried for years, but always failed to tell. Niagara is still,
and must always be, unpainted and unsung. It has flowed for
thousands of years as it thunders now, yet in its mighty rush
fresh beauties may be seen every hour, though its waters never
alter in their bulk for summer suns or the melting of Canadian snows.

www.ingramcontent.com/pod-product-compliance
Lightning Source LLC
Chambersburg PA
CBHW020328090426
42735CB00009B/1449